STUDENT WORKBOOK

Steven J. Peterson
Weber State University

Estimating in Building Construction

Seventh Edition

Frank R. Dagostino

Steven J. Peterson
Weber State University

Prentice Hall

Boston   Columbus   Indianapolis   New York   San Francisco   Upper Saddle River
Amsterdam   Cape Town   Dubai   London   Madrid   Milan   Munich   Paris   Montreal   Toronto
Delhi   Mexico City   Sao Paulo   Sydney   Hong Kong   Seoul   Singapore   Taipei   Tokyo

Editor in Chief: Vernon R. Anthony
Acquisitions Editor: David Ploskonka
Editorial Assistant: Nancy Kesterson
Director of Marketing: David Gesell
Executive Marketing Manager: Derril Trakalo
Senior Marketing Coordinator: Alicia Wozniak
Associate Managing Editor: Alexandrina Benedicto Wolf
Project Manager: Susan Hannahs
Cover Art: Shutterstock Images
Printer/Bindery: Bind-Rite Graphics
Cover Printer: Demand Production Center
Text Font: Times Roman
Credits and acknowledgments borrowed from other sources and reproduced, with permission, in this textbook appear on appropriate page within text.

Copyright © 2011 Pearson Education, Inc., publishing as Prentice Hall, One lake Street, Upper Saddle River, New Jersey, 07458. All rights reserved. Manufactured in the United States of America. This publication is protected by Copyright, and permission should be obtained from the publisher prior to any prohibited reproduction, storage in a retrieval system, or transmission in any form or by any means, electronic, mechanical, photocopying, recording, or likewise. To obtain permission(s) to use material from this work, please submit a written request to Pearson Education, Inc., Permissions Department, One Lake Street, Upper Saddle River, New Jersey, 07458.

Many of the designations by manufacturers and seller to distinguish their products are claimed as trademarks. Where those designations appear in this book, and the publisher was aware of a trademark claim, the designations have been printed in initial caps or all caps.

3 4 5 6 7 8 9 1 0   V088   14 13 12

www.pearsonhighered.com

ISBN 10: 0-13-509748-7

ISBN 13: 978-0-13-509748-9

Contents

Preface	v
Exercise 1: Introduction to Estimating	1
Exercise 2: Contracts, Bonds, and Insurance	3
Exercise 3: Project Manual	7
Exercise 4: The Estimate	11
Exercise 5: Computers in Estimating	15
Exercise 6: Overhead and Contingencies	17
Exercise 7: Labor Productivity and Labor Hours	19
Exercise 8: Labor Rates and Labor Cost	23
Exercise 9: Equipment	27
Exercise 10: Shrink and Swell	31
Exercise 11: Cross-Sectional Method	33
Exercise 12: Average End Area	45
Exercise 13: Area, Perimeter, and Topsoil	49
Exercise 14: Basement Excavation and Backfill	53
Exercise 15: Footing Excavation and Backfill	57
Exercise 16: Parking Lots	63
Exercise 17: Continuous Footings	65
Exercise 18: Spread Footings	69
Exercise 19: Piers and Columns	73
Exercise 20: Foundation Walls	77
Exercise 21: Grade Beams	83
Exercise 22: Slabs	85
Exercise 23: Masonry	89
Exercise 24: Metals	95
Exercise 25: Wood Framed Floors	101
Exercise 26: Wood Framed Walls	107
Exercise 27: Wood Framed Roofs	113iii
Exercise 28: Insulation	117
Exercise 29: Waterproofing and Roofing	123
Exercise 30: Doors and Windows	127
Exercise 31: Drywall and Plaster	131
Exercise 32: Flooring, Tile, and Paint	137
Exercise 33: Electrical	143
Exercise 34: Plumbing	145
Exercise 35: Heating, Ventilating, and Air-Conditioning	147
Exercise 36: Profit	149
Exercise 37: Other Estimating Methods	151

Preface

On my first day working in the construction industry, I was handed a set of plans and asked to prepare an estimate for the project. For me learning came by trial and error, with real bids and the company's profit margin at stake. You have the opportunity to learn to estimate in an environment where the stakes are not as high.

Estimating is an art that can only be learned through practice. This workbook has been developed to give the beginning estimator a foundation of experience on which to build his or her estimating career. This foundation can only be built through practicing correct estimating principles. The following are some things to keep in mind as you complete these problems:

1. Be sure to document all of the steps that you take to arrive at the solution. Where needed, add written statements explaining your estimate. You should be able to return to your estimates years from now and be able to explain how you reached your solution.

2. With each equation, be sure to include the units of measure and check to make sure that the units cancel out. A quantity is of little use without a unit of measure. Two hundred $2'' \times 8''$ @ $10'$ cannot be ordered without knowing if it is 200 lineal feet, 200 board feet, or 200 each. Canceling out units helps to ensure that the correct conversion factors are used. A large error can occur in an estimate when the wrong conversion factor is used. For example, using 9 square feet per square yard to convert from cubic feet to cubic yards overstates the needed concrete by 300 percent.

3. Double-check all of your math. Many estimating errors are the result of punching the wrong number into the calculator.

4. Many of the problems in this workbook occur in pairs (problems that are similar), which makes it tempting to skip a problem. Repetition is the key to remembering how to solve a problem. The best approach to developing a solid estimating foundation is to rework the example problems from a chapter in *Estimating in Building Construction*, working all the workbook problems that accompany the chapter, and then applying the estimating principals covered in the chapter on one or more of the drawings set that accompany *Estimating in Building Construction*.

Best of luck in your construction career.

Exercise 1: Introduction to Estimating

This exercise goes with Chapter 1 of *Estimating in Building Construction*.

1. For this assignment you will explore the role estimating plays in the construction industry by interviewing a person whose job duties include estimating. Begin by setting up an interview with an estimator, project manager, project engineer, superintendent, foreperson, architect, engineer, construction material salesperson, or freelance estimator. During the interview, ask the person the following questions and ask follow-up questions as necessary. Be respectful of their time and limit your interview to 20 minutes, unless the person offers to extend the interview. Be sure to thank the person before you leave and mail them a thank you note within 48 hours of the interview. After the interview, prepare written responses to the following questions and be prepared to discuss your findings in class, if your instructor chooses to do so:

    a. What are the estimates used for (ordering materials, preliminary budget, etc.)?

    b. At what stage of the construction process (early-design, late-design, bidding, construction, etc.) does the estimate occur?

    c. What are the consequences if the estimate is slightly wrong? If it is very wrong?

    d. How do they prepare an estimate? After the interview, decide which estimating method (detailed, assembly, square-foot, parametric, model, or project comparison) best describes the type of estimates he or she prepared.

    e. How long does it take to prepare an estimate?

    f. What skills are required to become a good estimator?

    g. What experience is required to get a job like his or hers?

2. Obtain a copy of the contract documents (drawings and project manual) for a construction project. Contractors will often give you an unneeded set of contract documents after the project is complete. For some projects, the contract documents may be downloaded from the Internet. Write a brief summary of how the contract documents are organized. Be sure to discuss both the project manual and the drawings. Be prepared to discuss your findings in class, if your instructor chooses to do so.

3. Using the Warehouse.xls Excel file that accompanies *Estimating in Building Construction*, determine the estimated cost of a warehouse with the following parameters:
    Building length—210 feet
    Number of bays on the length side of the building—7 each
    Building width—120 feet
    Number of bays on the width of the building—4 each
    Wall height above grade—22 feet
    Depth to top of footing—12 inches
    Floor slab—6 inch thick with wire mesh
    Number of roof hatches—2 each
    Number of personnel doors—4 each
    Number of 14-foot-wide by 14-foot-high overhead doors—14 each
    Number of 4-foot by 4-foot skylights—28 each
    Fire sprinklers are not required
    Separate male and female bathrooms are required

Exercise 2: Contracts, Bonds, and Insurance

This exercise goes with Chapter 2 of *Estimating in Building Construction.*

Answer the following questions using the contract documents (drawings and project manual) obtained for Exercise 1, Problem 2 of this workbook. Be prepared to discuss your findings in class, if your instructor chooses to do so.

1. What type of agreement (lump-sum, unit price, or cost-plus-fee) is used for the project? If it is a cost-plus-fee agreement, how is the fee determined, and is there a guaranteed maximum price?

2. What is the scope of the work for the project?

3. What provisions are included in the contract documents regarding the time of completion? What penalties are there for failing to meet the completion date? Is there a bonus for completing the project ahead of schedule?

4. How are progress payments handled? When are they due? How quickly will they be paid?

5. Will retention be withheld? If so, how much? What are the requirements for the release of retention?

6. How is final acceptance handled? What inspections are required? What forms, documents, maintenance and operation manuals, certifications, red-line drawings, etc., need to be submitted before final acceptance?

7. What bonds are needed for the job?

8. What are the insurance requirements for the project?

Exercise 3: Project Manual

This exercise goes with Chapter 3 of *Estimating in Building Construction*.

Answer the following questions using the contract documents (drawings and project manual) obtained for Exercise 1, Problem 2 of this workbook. Be prepared to discuss your findings in class, if your instructor chooses to do so.

1. How can a contractor obtain a set of contract documents?

2. Are the contract documents on file at plan rooms, trade associations, or government agencies? If so, where are they on file?

3. When and where is the bid due?

4. What needs to be submitted for the bid to be complete?

5. Do the contract documents use a standard contract prepared by the American Institute of Architects, a trade association, or government agency? If so, who prepared the contract?

6. Do the contract documents use a standard set of general conditions prepared by the American Institute of Architects, a trade association, or government agency? If so, who prepared the general conditions?

7. Do the contract documents contain supplementary general conditions? If so, how do they modify the general conditions?

8. What alternatives need to be included in the bid?

9. Have any addendums been issued for the project? Is so, how do they modify or clarify the contract documents?

10. What allowances, if any, need to be included in the bid?

11. Select one section from the technical specifications and answer the following questions. Be sure to select a section that deals with specific components of the building (e.g., doors, concrete, masonry, etc.).

a. What scope of work does this section cover? What is specifically included? What is specifically excluded?

b. What submittals, if any, are required under this section?

c. How does this section address quality?

d. What materials are required to complete the work under this section?

e. What installation procedures are required by this section?

f. What warranties are required for this section?

g. What else is contained in this section?

h. What did you find interesting about this section?

Exercise 4: The Estimate

This exercise goes with Chapter 4 of *Estimating in Building Construction*.

1. Using the contract documents (drawings and project manual) obtained for Exercise 1, Problem 2 of this workbook, discuss the pros and cons of bidding on the project. Be prepared to present your findings in class, if your instructor chooses to do so.

    Pros:

    Cons:

2. You are working as an estimator for a small general contractor. The company has its own finish carpentry, framing, and concrete crews. The finish carpentry crew can perform all types of finish carpentry. The framing and concrete crews can perform typical framing (both wood and metal studs) and typical concrete work, respectively. The framing and concrete crews are not capable of performing complex or specialized framing or concrete work. Using the contract documents (drawings and project manual) obtained for Exercise 1, Problem 2 of this workbook, answer the following questions. Be prepared to present your findings in class, if your instructor chooses to do so.

    a. What part of the work would be performed by your company's crews? How many hours do you think it will take you to bid this work?

    b. From what material suppliers will you need prices to complete the bid?

c. From what subcontractors will you need prices to complete the bid?

d. What else do you need to complete the bid?

e. How many days will it take you to prepare the entire bid, including getting material quotes and subcontractor bids?

3. Identify one potential supplier for each of the suppliers identified in Problem 2b.

4. Identify one potential subcontractor for each of the subcontractors identified in Problem 2c.

Exercise 5: Computers in Estimating

This exercise goes with Chapter 5 of *Estimating in Building Construction.*

1. Select two specialized estimating software packages. How do the features of the packages compare? Which package would you choose, and why? Be prepared to present your findings in class, if your instructor chooses to do so.

2. Select a takeoff software package. Discuss the advantages and disadvantages of using the selected software package compared to performing takeoffs from a set of paper plans. Be prepared to present your findings in class, if your instructor chooses to do so.

Exercise 6: Overhead and Contingencies

This exercise goes with Chapter 6 of *Estimating in Building Construction*.

1. Using the contract documents (drawings and project manual) obtained for Exercise 1, Problem 2 of this workbook, prepare a preliminary schedule for the project assuming the project is to begin April 1 of this year. The schedule should be of sufficient detail to determine the duration of the project and the overhead budget. Be prepared to present your schedule in class, if your instructor chooses to do so.

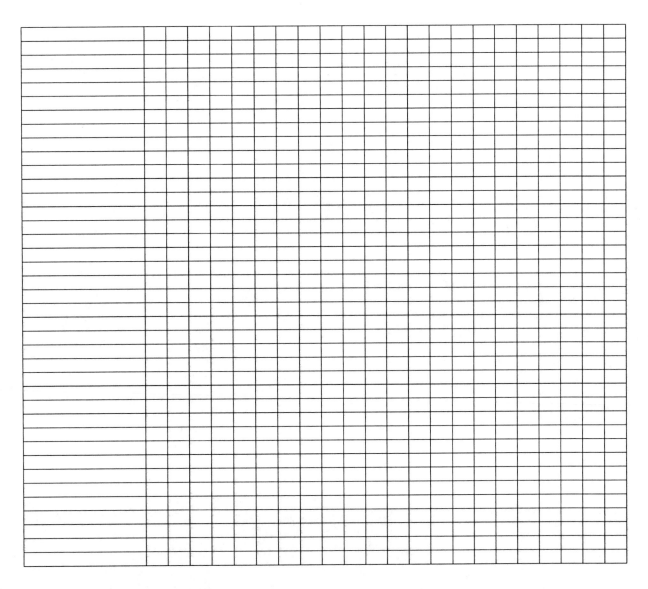

2. Using the contract documents (drawings and project manual) obtained for Exercise 1, Problem 2 of this workbook, prepare a list of overhead items needed to complete your project. Identify when and how long each item will be needed during the project. For example, the mobilization and demobilization of the temporary office will be required at the beginning and end of the project, and the temporary office will be required during the entire project. Use your schedule from Problem 1 of this exercise to help you determine when overhead items are needed. Be prepared to present your list in class, if your instructor chooses to do so.

Exercise 7: Labor Productivity and Labor Hours

This exercise goes with Chapter 7 of *Estimating in Building Construction*.

1. During the construction of a recent residence, your company's finish carpentry crew installed 14 interior doors in 18 labor hours. Determine productivity rate in labor hours per door for installing doors on this project.

2. Using the productivity from Problem 1, how many hours would it take to install 27 interior doors?

3. During the construction of a recent office building, your company's framing crew installed 515 lineal feet of 8-foot-high, metal-stud wall in two days. The crew consisted of two carpenters and one helper and worked eight hours per day. Determine productivity rate in labor hours per foot of wall for installing metal-stud walls on this project.

4. Using the productivity from Problem 3, determine the number of labor hours to install 810 lineal feet of 8-foot-high, metal-stud wall. If the crew consisted of a carpenter and laborer, how many days would it take to frame the wall? They work eight hours per day.

5. A commercial project requires the installation of 1,856 square feet of acoustical ceiling. Using a productivity rate of 0.013 labor hours per square foot and a productivity factor of 0.95, determine the number of labor hours required to install the ceiling.

6. A commercial project requires the installation of 127 square yards of carpet. Using a productivity rate of 0.108 labor hours per square yard and a productivity factor of 1.05, determine the number of labor hours required to install the carpet.

7. A number of dump trucks are used to haul asphalt from the plant to a road construction project. The estimated cycle time for the trucks is 55 minutes. The trucks carry 15 tons of asphalt per trip. Using a system efficiency of 50 minutes per hour and productivity factor of 1.1, determine the number of labor hours required to haul one ton of asphalt. Each truck requires one operator.

8. Scrapers are used to haul dirt from a borrow pit to the cap of a landfill. The estimated cycle time for the scrapers is 9.5 minutes. The scrapers carry 10 cubic yards of soil per trip. Using a system efficiency of 45 minutes per hour and productivity factor of 1.0, determine the number of labor hours to haul 1 cubic yard of dirt. Each scraper requires one operator.

9. An asphalt paving crew is used to pave a road 8,000 feet long. The road is 48 feet wide and the paving machine can pave a 12-foot-wide strip of road during each pass. The machine paves 13 lineal feet of 12-foot-wide road per minute. Two lifts of asphalt are needed to pave the road. After completing a pass, it takes the crew 15 minutes to turn the paving machine around and get ready for the next pass. The crew consists of a foreperson, seven operators, and a laborer. Using a system efficiency of 45 minutes per hour, determine the number of labor hours required to place 1 lineal foot of road 12 feet wide.

10. An asphalt paving crew is used to pave a 500-foot by 300-foot parking lot. The parking lot will be paved in 12-foot-wide strips running the long (500-foot) direction. The machine paves 11 lineal feet of 12-foot-wide parking lot per minute. A single lift of asphalt will be used on the parking lot. After completing a pass, it takes the crew 10 minutes to turn the paving machine around and get ready for the next pass. The crew consists of a foreperson, six operators, and a laborer. Using a system efficiency of 50 minutes per hour, determine the number of labor hours required to place 1 lineal foot of parking lot 12 feet wide.

Exercise 8: Labor Rates and Labor Cost

This exercise goes with Chapter 7 of *Estimating in Building Construction*.

1. Given the following information, determine the burdened hourly wage rate for a carpenter. Assume the carpenters take full advantage of the retirement benefit.
    Wage rate—$23.00 per hour
    Hours worked—45 hours per week for 30 weeks, and 40 hours per week for 20 weeks
    Vacation, holidays, and sick leave—Two weeks at 40 hours per week
    Overtime—Time-and-a-half for any hours over 40 per week
    Gas allowance—$150 per month
    Annual bonus—$300
    Social Security—6.2 percent on the first $106,800 of wages
    Medicare—1.45 percent of all wages
    FUTA—0.8 percent on the first $7,000 of wages
    SUTA—5.8 percent on the first $10,000 of wages
    Worker's compensation insurance—$12.25 per $100.00 of wages
    General liability insurance—0.55 percent of wages
    Health insurance (company's portion)—$450 per month per employee
    Retirement—$0.50 per $1.00 contributed by the employee on 6 percent of the employee's wages

2. Given the following information, determine the burdened hourly wage rate for an equipment operator. Assume the equipment operators take full advantage of the retirement benefit.
    Wage rate—$36.25 per hour
    Hours worked—50 hours per week for 40 weeks, and 40 hours per week for nine weeks
    Vacation, holidays, and sick leave—Three weeks at 40 hours per week
    Overtime—Time-and-a-half for any hours over 40 per week
    Gas allowance—$100 per month
    Annual bonus—$750
    Social Security—6.2 percent on the first $106,800 of wages
    Medicare—1.45 percent of all wages
    FUTA—0.8 percent on the first $7,000 of wages
    SUTA—2.5 percent on the first $22,000 of wages
    Worker's compensation insurance—$9.27 per $100.00 of wages
    General liability insurance—0.65 percent of wages
    Health insurance (company's portion)—$385 per month per employee
    Retirement—$0.75 per $1.00 contributed by the employee on 6 percent of the employee's wages

3. Given the following information, determine the burdened monthly wage rate for a project manager. Assume the project manager takes full advantage of the retirement benefit.

    Wage rate—$8,800 per month
    Gas allowance—$200 per month
    Annual bonus—$5,000
    Social Security—6.2 percent on the first $106,800 of wages
    Medicare—1.45 percent of all wages
    FUTA—0.8 percent on the first $7,000 of wages
    SUTA—3.1 percent on the first $18,000 of wages
    Worker's compensation insurance—$6.60 per $100.00 of wages
    General liability insurance—0.8 percent of wages
    Health insurance (company's portion)—$400 per month per employee
    Retirement—$0.50 per $1.00 contributed by the employee on 6 percent of the employee's wages

4. It is estimated that it will take 217 labor hours to construct a block wall. Using a burdened hourly rate of $38.75 per labor hour, determine the labor cost to construct the wall.

5. It is estimated that it will take 538 labor hours to hang drywall in a commercial building. Using a burdened hourly rate of $27.49 per labor hour, determine the labor cost to hang the drywall.

Exercise 9: Equipment

This exercise goes with Chapter 8 of *Estimating in Building Construction*.

1. Using the following information, determine the fixed and operating costs for a scraper.
    Actual cost (delivered)—$156,000
    Horsepower rating—200 hp
    Cost of tires—$6,000
    Salvage value—20 percent
    Useful life—6 years (8,400 hours)
    Total interest—7 percent
    Length of loan—6 years
    Total insurance, taxes, and storage—5.5 percent per year
    Fuel cost—$3.90 per gallon
    Fuel consumption rate—0.06 gallons per hp per hour
    Power utilization—85 percent
    Use factor—40 minutes per hour (66.67 percent)
    Lubrication—8 quarts of oil at $4.50 per quart
    Oiler labor—2.2 hours at $27.65
    Lubrication schedule—Every 100 hours
    Life of tires—3,500 hours
    Repair to tires—12 percent of deprecation
    Repairs to equipment—120 percent over useful life

2. Using the following information, determine the fixed and operating costs for a hydraulic excavator.
    Actual cost (delivered)—$172,600
    Horsepower rating—245 hp
    Salvage value—10 percent
    Useful life—8 years (12,800 hours)
    Total interest—6.5 percent
    Length of loan—8 years
    Total insurance, taxes, and storage—4.5 percent per year
    Fuel cost—$3.65 per gallon
    Fuel consumption rate—0.04 gallons per hp per hour
    Power utilization—60 percent
    Use factor—50 minutes per hour (83.33 percent)
    Lubrication—7 quarts of oil at $4.65 per quart
    Oiler labor—2.9 hours at $31.27
    Lubrication schedule—Every 150 hours
    Repairs to equipment—75 percent over useful life

3. Visit a local equipment supplier. Find out what information they can provide about the ownership and operation costs of their equipment. Prepare a written summary of your findings. Be prepared to present your findings in class, if your instructor chooses to do so.

Exercise 10: Shrink and Swell

This exercise goes with Chapter 9 of *Estimating in Building Construction*.

1. For a project, your company needs to haul away 1,450 bank cubic yards (bcy) of soil. If the soil has a swell of 25 percent, how many loose cubic yards (lcy) of soil will need to be hauled? How many truckloads are required to haul the soil offsite if each truck can haul nine loose cubic yards?

2. For a project, your company needs to haul away 350 bank cubic yards of sand. If the sand has a swell of 12 percent, how many loose cubic yards of sand will need to be hauled? How many truckloads are required to haul the sand offsite if each truck can haul 12 loose cubic yards?

3. If 750 compacted cubic yards (ccy) of in-place soil is required for a project, how many loads of import will be required? The import material has a swell of 30 percent and shrinkage of 90 percent. The trucks can haul 10 loose cubic yards.

4. If 1,490 compacted cubic yards of in-place soil is required for a project, how many loads of import will be required? The import material has a swell of 14 percent and shrinkage of 95 percent. The trucks can haul 12 loose cubic yards.

Exercise 11: Cross-Sectional Method

This exercise goes with Chapter 9 of *Estimating in Building Construction*.

1. Using the cross-sectional method, determine the cuts and fill for the project shown in Figure 11-1. The grids are 50 feet apart in both directions. The existing grade appears above the proposed grade. Express your cuts in bank cubic yards (bcy) and your fills in compacted cubic yards (ccy).

	A	B	C	D	E	F	G	
1	104.5 / 104.5	104.7 / 104.7	105.1 / 105.1	105.3 / 105.3	105.9 / 105.9	106.1 / 106.1	106.5 / 106.5	1
		1	2	3	4	5	6	
2	104.2 / 104.2	104.6 / 104.5	104.8 / 104.5	105.4 / 104.5	105.6 / 104.5	106.0 / 104.5	106.0 / 106.0	2
		7	8	9	10	11	12	
3	103.9 / 103.9	104.2 / 104.5	104.7 / 104.5	104.9 / 104.5	105.2 / 104.5	105.4 / 104.5	105.5 / 105.5	3
		13	14	15	16	17	18	
4	103.4 / 103.4	103.8 / 104.5	104.4 / 104.5	104.8 / 104.5	105.1 / 104.5	105.1 / 104.5	105.3 / 105.3	4
		19	20	21	22	23	24	
5	103.1 / 103.1	103.3 / 104.5	103.6 / 104.5	104.0 / 104.5	104.6 / 104.5	104.7 / 104.5	105.0 / 105.0	5
		25	26	27	28	29	30	
6	102.8 / 102.8	103.2 / 104.5	103.6 / 104.5	103.9 / 104.5	104.0 / 104.5	103.7 / 104.5	103.9 / 103.9	6
		31	32	33	34	35	36	
7	102.7 / 102.7	102.9 / 102.9	103.1 / 103.1	103.5 / 103.5	103.9 / 103.9	103.4 / 103.4	103.7 / 103.7	7
	A	B	C	D	E	F	G	

Figure 11   Grades for Problem 11-1

Cell 1:

Cell 2:

Cell 3:

Cell 4:

Cell 5:

Cell 6:

Cell 7:

Cell 8:

Cell 9:

Cell 10:

Cell 11:

Cell 12:

Cell 13:

Student Workbook to Accompany Estimating in Building Construction 37

Cell 14:

Cell 15:

Cell 16:

Cell 17:

Cell 18:

Cell 19:

Cell 20:

Cell 21:

Cell 22:

Cell 23:

Cell 24:

Cell 25:

Cell 26:

Cell 27:

Cell 28:

Cell 29:

Cell 30:

Cell 31:

Cell 32:

Cell 33:

Cell 34:

Cell 35:

Cell 36:

Total Cuts:

Total Fills:

2. Set up Problem 1 in a spreadsheet.

3. The material in Problem 1 has a shrinkage of 95 percent and a swell of 30 percent. Determine the net import or export required for the project and the number of 9-cubic-yard truckloads needed to haul the material to or from the project.

Exercise 12: Average End Area

This exercise goes with Chapter 9 of *Estimating in Building Construction*.

1. Using the average end area method, determine the required volume of the cuts and fills for a new road. The cuts and fills for each station are shown in Table 12-1. Express your answer in cubic yards.

Table 12-1
Cuts and fills for Problem 1

Station	Cut (sf)	Fill (sf)
10 + 00	0	0
10 + 50	257	0
11 + 00	175	50
11 + 50	20	285
12 + 00	0	233
12 + 20	0	0

Cuts:

Fills:

2. The material in Problem 1 has a shrinkage of 97 percent and a swell of 18 percent. Determine the net import or export required for the project and the number of 11-cubic-yard truckloads needed to haul the material to or from the project.

3. Using the average end area method, determine the required volume of the cuts and fills for a new road. The cuts and fills for each station are shown in Table 12-2. Express your answer in cubic yards.

Table 12-2
Cuts and fills for Problem 3

Station	Cut (sf)	Fill (sf)
8 + 30	0	0
8 + 50	125	25
9 + 00	168	0
9 + 50	210	0
10 + 00	144	95
10 + 50	75	65
10 + 75	0	0

Cuts:

Fills:

4. The material in Problem 3 has a shrinkage of 94 percent and a swell of 25 percent. Determine the net import or export required for the project and the number of 10-cubic-yard truckloads needed to haul the material to or from the project.

Exercise 13: Area, Perimeter, and Topsoil

This exercise goes with Chapter 9 of *Estimating in Building Construction*.

1. Find the perimeter of the building in Figure 13-1.

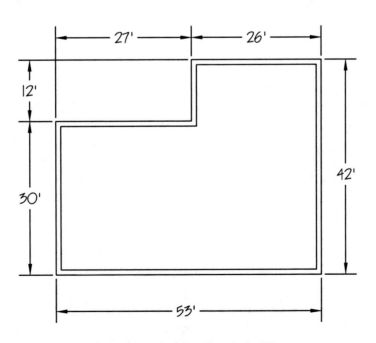

Figure 13-1 Building Foundation Plan

2. Find the area of the building in Figure 13-1.

3. Six inches of topsoil is to be removed and stockpiled for the building in Figure 13-1. The topsoil is to be removed 8 feet back from the perimeter of the building to allow for working around the building and sloping the sides of the excavation. How many bank cubic yards (bcy) of topsoil need to be removed and stockpiled?

4. The topsoil in Problem 3 is to be removed with a bulldozer. The estimated productivity is 22 bcy per hour and it will take two hours to mobilize the bulldozer. The total ownership and operation cost of the bulldozer is $42.56 per hour and the cost of an operator is $22.14 per hour. Determine the number of hours and the cost to strip the topsoil.

5. Eight inches of topsoil is to be removed and stockpiled for the building in Figure 13-2. The topsoil is to be removed 10 feet back from the perimeter of the building to allow for working around the building and sloping the sides of the excavation. How many bank cubic yards of topsoil need to be removed and stockpiled?

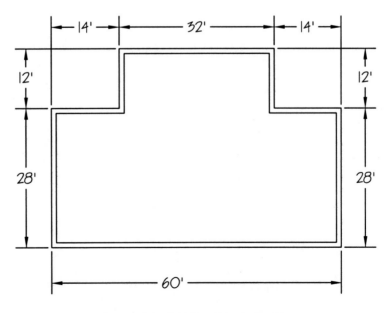

Figure 13-2 Building Foundation Plan

6. The topsoil in Problem 5 is to be removed with a bulldozer. The estimated productivity is 30 bcy per hour and it will take 2.5 hours to mobilize the bulldozer. The total ownership and operation cost of the bulldozer is $47.12 per hour and the cost of an operator is $26.88 per hour. Determine the number of hours and the cost to strip the topsoil.

Exercise 14: Basement Excavation and Backfill

This exercise goes with Chapter 9 of *Estimating in Building Construction*.

1. Determine the amount of excavation needed for the basement in Figures 14-1 and 14-2. A two-foot working distance is required and the excavation will be sloped 1.5:1 (1.5 vertical feet for every horizontal foot).

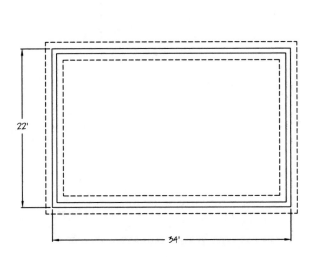

Figure 14-1 Basement Plan View

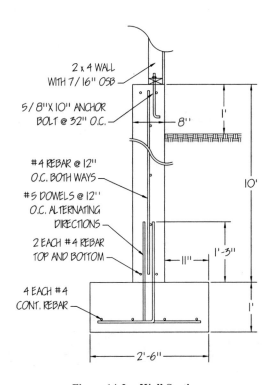

Figure 14-2 Wall Section

2. Determine the amount of backfill needed for the basement in Problem 1. Ignore any backfill inside of the walls.

3. Determine the amount of excavation needed for the basement in Figures 14-3 and 14-4. A one-foot working distance is required and the excavation will be sloped 2:1 (two vertical feet for every horizontal foot). *Hint:* Calculate the excavation for each of the areas separately and add them together.

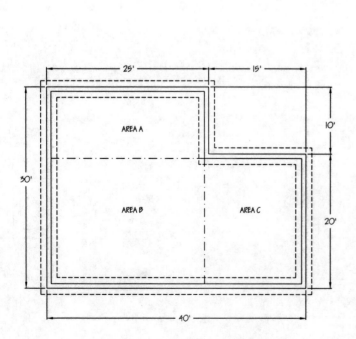

Figure 14-3   Basement Plan View

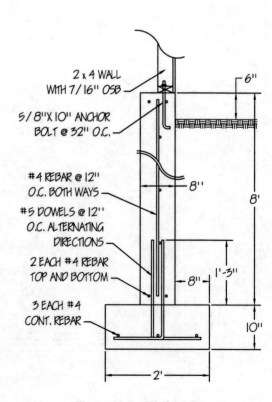

Figure 14-4   Wall Section

4. Determine the amount of backfill needed for the basement in Problem 3. Ignore any backfill inside of the walls.

Exercise 15: Footing Excavation and Backfill

This exercise goes with Chapter 9 of *Estimating in Building Construction*.

1. Determine the amount of excavation needed for the continuous footings in Figures 15-1 and 15-2. A one-foot working distance is required and the excavation will be sloped 1:1 (one vertical foot for every horizontal foot).

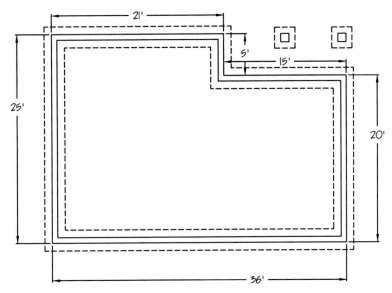

Figure 15-1   Foundation Plan View

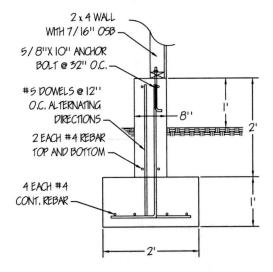

Figure 15-2   Wall Section

2. Determine the amount of backfill needed for the continuous footing and foundation in Problem 1.

3. Determine the amount of excavation needed for the spread footings in Figures 15-1 and 15-3. A one-foot working distance is required and the excavation will be sloped 1:1 (one vertical foot for every horizontal foot).

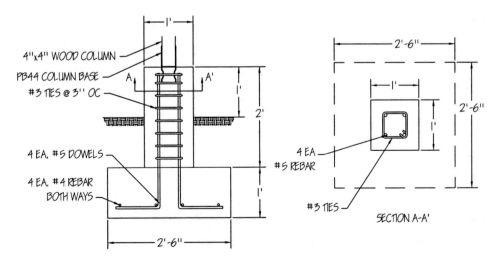

Figure 15-3   Column Section and Plan View

4. Determine the amount of backfill needed for the spread footings and columns in Problem 3.

5. Determine the amount of excavation needed for the continuous footings in Figures 15-4 and 15-5. A two-foot working distance is required and the excavation will be sloped 1:1.75 (1.75 vertical feet for every horizontal foot).

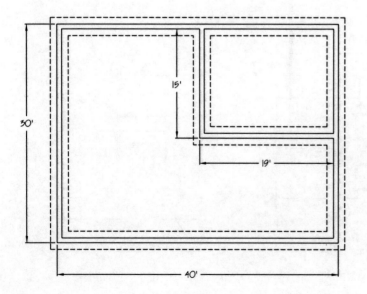

Figure 15-4   Foundation Plan View

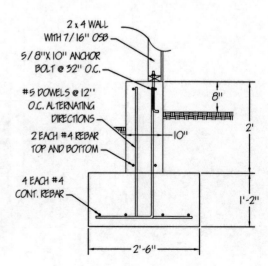

Figure 15-5   Wall Section

6. Determine the amount of backfill needed for the continuous footing and foundation in Problem 5.

7. Determine the amount of excavation needed for the spread footing in Figure 15-6. A three-foot working distance is required and the excavation will be sloped 1:1 (one vertical foot for every horizontal foot).

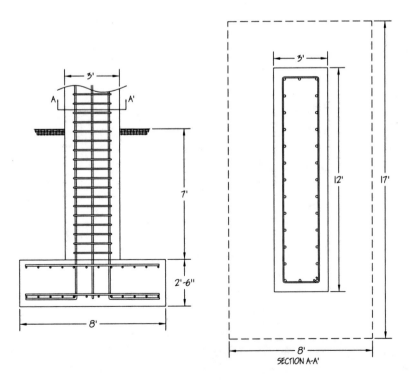

Figure 15-6 Column Section and Plan View

8. Determine the amount of backfill needed for the spread footing and column in Problem 7.

Exercise 16: Parking Lots

This exercise goes with Chapter 9 of *Estimating in Building Construction*.

1. Determine the quantity of asphalt and granular material required for the parking lot shown in Figure 16-1. The asphalt is 4 inches thick and the granular material is 8 inches thick.

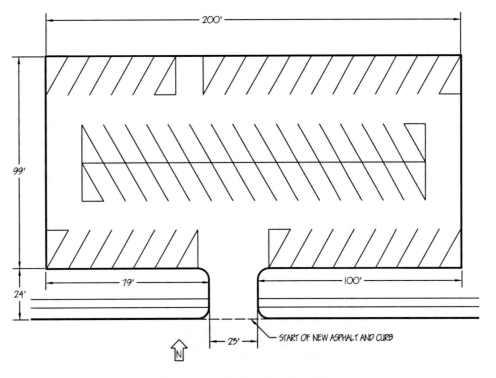

Figure 16-1 Parking Lot Plan View

2. Determine the quantity of asphalt and granular material required for the parking lot shown in Figure 16-2. The asphalt is 3 inches thick and the granular material is 6 inches thick.

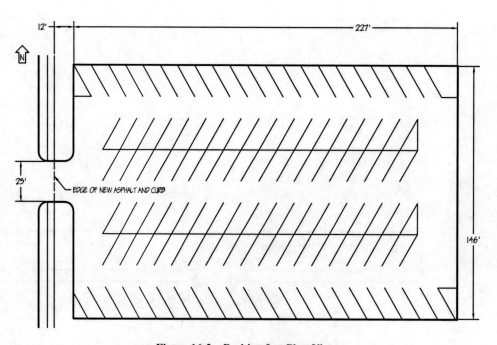

Figure 16-2 Parking Lot Plan View

Exercise 17: Continuous Footings

This exercise goes with Chapter 10 of *Estimating in Building Construction*.

1. Using a waste factor of 7 percent, determine the number of cubic yards of concrete needed to pour the continuous footings shown in Figures 17-1 and 17-2. Assume that the wall is centered over the footing.

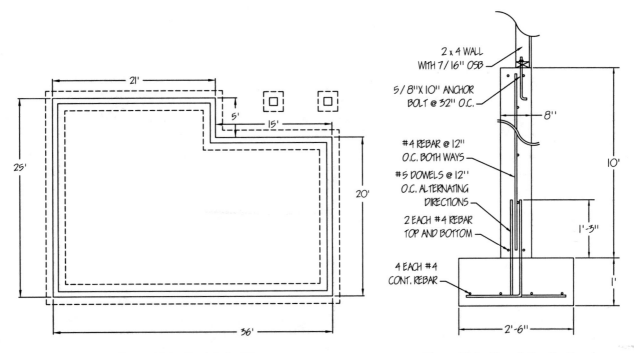

Figure 17-1 Foundation Plan   Figure 17-2 Foundation Section

65

2. Determine the amount of rebar needed for the continuous footing shown in Figures 17-1 and 17-2. Allow for two inches of cover and extend the dowels 15 inches into the wall. Add 10 percent for lap and waste to the continuous bars.

3. Determine the number of linear feet of forms needed for the continuous footings shown in Figures 17-1 and 17-2.

4. Using a waste factor of 8 percent, determine the number of cubic yards of concrete needed to pour the continuous footings shown in Figures 17-3 and 17-4. Assume that the wall is centered over the footing.

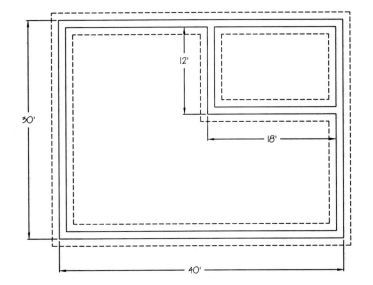

Figure 17-3 Foundation Plan

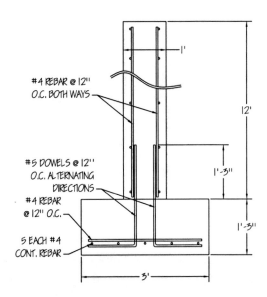

Figure 17-4 Foundation Section

5. Determine the amount of rebar needed for the continuous footing shown in Figures 17-3 and 17-4. Allow for two inches of cover and extend the dowels 15 inches into the wall. Add 10 percent for lap and waste to the continuous bars.

6. Determine the number of linear feet of forms needed for the continuous footings shown in Figures 17-3 and 17-4.

Exercise 18: Spread Footings

This exercise goes with Chapter 10 of *Estimating in Building Construction*.

1. Using a waste factor of 8 percent, determine the number of cubic yards of concrete needed to pour two of the spread footings in Figure 18-1.

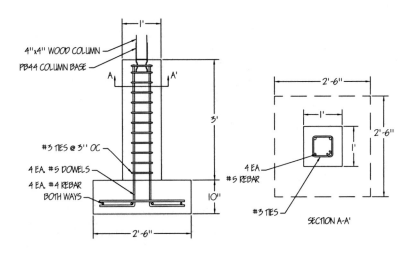

Figure 18-1   Column Section and Plan View

2. Determine the amount of rebar needed for two of the spread footings shown in Figure 18-1. Allow for two inches of cover.

3. Using a waste factor of 10 percent, determine the number of cubic yards of concrete needed to pour the spread footing in Figure 18-2.

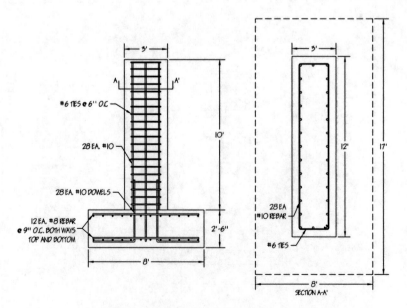

Figure 18-2 Column Section and Plan View

4. Determine the amount of rebar needed for the spread footing Figure 18-2. The dowels extend 24 inches into the column. Allow for three inches of cover.

Exercise 19: Piers and Columns

This exercise goes with Chapter 10 of *Estimating in Building Construction*.

1. Using a waste factor of 5 percent, determine the number of cubic yards of concrete needed to pour two of the piers in Figure 19-1.

2. Determine the amount of rebar needed for two of the piers shown in Figure 19-1. Allow for two inches of cover.

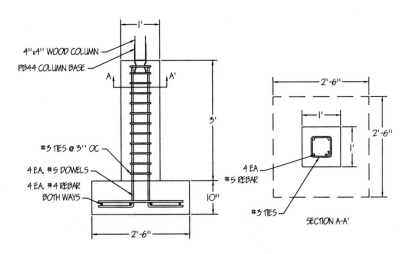

Figure 19-1 Column Section and Plan View

3. Using a waste factor of 4 percent, determine the number of cubic yards of concrete needed to pour the column in Figure 19-2.

4. Determine the amount of rebar needed for the column shown in Figure 19-2. Allow for three inches of cover.

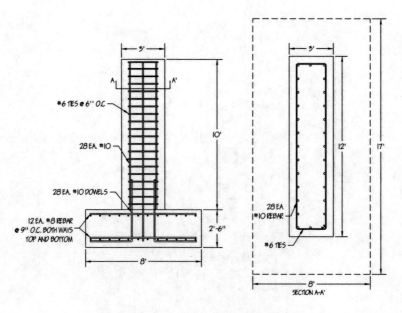

Figure 19-2 Column Section and Plan View

5. A building requires four drilled piers with a shaft diameter of 24 inches, a bell diameter of 48 inches, and a bell angle of 50 degrees. The height from the bottom of the bell to the top of the pier is 10 feet. Using a waste factor of 12 percent, determine the number of cubic yards of concrete needed to pour the drilled piers.

6. A building requires 10 drilled piers with a shaft diameter of 30 inches, a bell diameter of 54 inches, and a bell angle of 60 degrees. The height from the bottom of the bell to the top of the pier is 15 feet. Using a waste factor of 15 percent, determine the number of cubic yards of concrete needed to pour the drilled piers.

Exercise 20: Foundation Walls

This exercise goes with Chapter 10 of *Estimating in Building Construction*.

1. Using a waste factor of 5 percent, determine the number of cubic yards of concrete needed to pour the foundation walls shown in Figures 20-1 and 20-2.

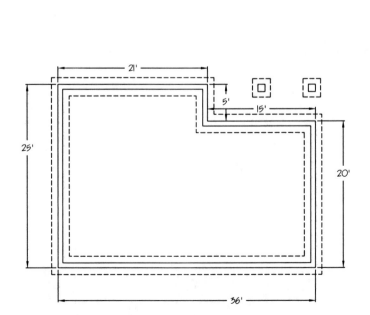

Figure 20-1   Foundation Plan

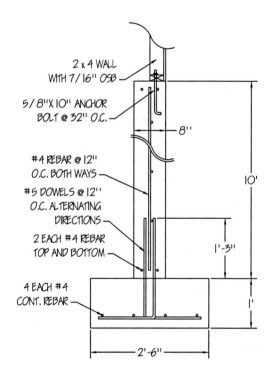

Figure 20-2   Foundation Section

2. Determine the amount of rebar needed for the foundation walls shown in Figures 20-1 and 20-2. Allow for two inches of cover. Add 10 percent for lap and waste tzo the continuous bars.

3. Determine the cost to install the rebar for the foundation walls in Figures 20-1 and 20-2 using a productivity of 10.75 labor hours per ton and an average labor rate of $25.62 per labor hour.

4. Determine the square foot contact area (SFCA) of the forms needed for the foundation walls shown in Figures 20-1 and 20-2.

5. Determine the cost to form the foundation walls in Figures 20-1 and 20-2 using a productivity of 0.125 labor hours per SFCA and an average labor rate of $25.62 per labor hour.

6. Using a waste factor of 6 percent, determine the number of cubic yards of concrete needed to pour the foundation walls shown in Figures 20-3 and 20-4.

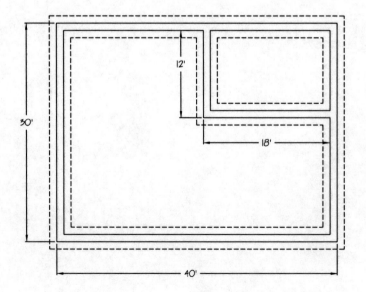

Figure 20-3   Foundation Plan

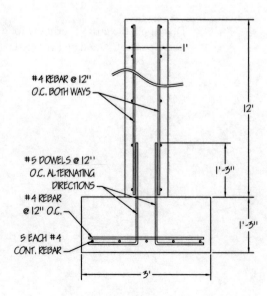

Figure 20-4   Foundation Section

7. Determine the amount of rebar needed for the foundation walls shown in Figures 20-3 and 20-4. Allow for two inches of cover. Add 10 percent for lap and waste to the continuous bars.

8. Determine the cost to install the rebar for the foundation walls in Figures 20-3 and 20-4 using a productivity of 10.3 labor hours per ton and an average labor rate of $31.26 per labor hour.

9. Determine the square foot contact area (SFCA) of the forms needed for the foundation walls shown in Figures 20-3 and 20-4.

10. Determine the cost to form the foundation walls in Figures 20-3 and 20-4 using a productivity of 0.11 labor hours per SFCA and an average labor rate of $32.17 per labor hour.

Exercise 21: Grade Beams

This exercise goes with Chapter 10 of *Estimating in Building Construction*.

1. A building requires 20 feet of the grade beam whose cross section is shown in Figure 21-1. Using a waste factor of 8 percent, determine the number of cubic yards of concrete needed to pour the grade beam.

2. Determine the amount of rebar needed for the grade beam in Problem 1. Allow for two inches of cover and ignore waste.

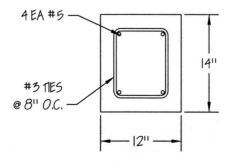

Figure 21-1   Grade Beam Section

3. A building requires 35 feet of the grade beam whose cross section is shown in Figure 21-2. Using a waste factor of 10 percent, determine the number of cubic yards of concrete needed to pour the grade beam.

4. Determine the amount of rebar needed for the grade beam in Problem 3. Allow for three inches of cover. Add 10 percent for lap and waste to the continuous bars.

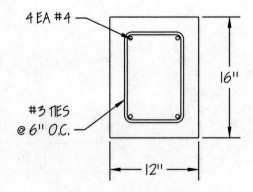

Figure 21-2   Grade Beam Section

Exercise 22: Slabs

This exercise goes with Chapter 10 of *Estimating in Building Construction*.

1. A building requires a 100-foot-long by 83-foot-wide by 5-inch-thick concrete slab. Using a waste factor of 10 percent, determine the number of cubic yards of concrete needed to pour the slab.

2. Determine the number of rolls of wire mesh needed for the slab in Problem 1. There are 750 square feet of wire mesh per roll. Add 12 percent for lap and waste.

3. Determine the amount of rebar needed for the slab in Problem 1. The slab is reinforced with #4 rebar at 12 inches on center both ways. Assume two inches of cover and add 10 percent for lap and waste.

4. Determine the cost to finish the slab in Problem 1 using a productivity of 0.011 labor hours per square foot and an average labor rate of $29.92 per labor hour.

5. A vapor barrier is to be placed below the slab in Problem 1. Determine the number of 10-foot by 100-foot rolls of vapor barrier needed. The vapor barrier needs to be lapped one foot. *Hint:* Subtract the lap from the width of the roll.

6. Using a waste factor of 12 percent, determine the amount of concrete needed to pour the slab shown in Figure 22-1. The slab is 4 inches thick.

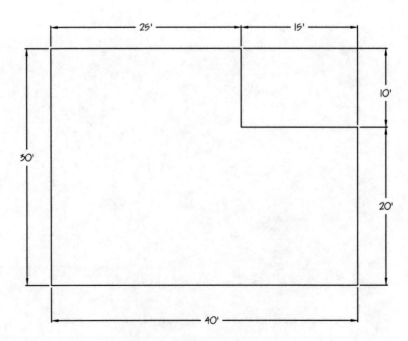

Figure 22-1 Floor Slab Perimeter

7. The slab in Problem 6 is reinforced with wire mesh. Determine the number of rolls of wire mesh needed for the slab. There are 750 square feet of wire mesh per roll. Add 11 percent for lap and waste.

8. Determine the amount of rebar needed for the slab in Problem 6. The slab is reinforced with #4 rebar at 18 inches on center both ways. Assume two inches of cover and add 10 percent for lap and waste.

9. Determine the cost to finish the slab in Problem 6 using a productivity of 0.013 labor hours per square foot and an average labor rate of $31.17 per labor hour.

10. A vapor barrier is to be placed below the slab in Problem 6. Determine the number of 10-foot by 50-foot rolls of vapor barrier needed. The vapor barrier needs to be lapped one foot.

Exercise 23: Masonry

This exercise goes with Chapter 11 of *Estimating in Building Construction*.

1. Determine the number of 8-inch-high by 8-inch-wide by 16-inch-long concrete (CMU) blocks required to complete 250 feet of the fence whose cross section is shown in Figure 23-1. If lintel blocks are required wherever the #4 horizontal bars are located, how many plain blocks and how many lintel blocks are needed for the wall?

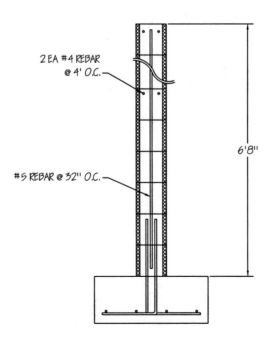

Figure 23-1 Wall Section

2. Determine the amount of rebar needed for the CMU fence in Problem 1. Allow for two inches of cover. Add 10 percent for lap and waste to the horizontal bars.

3. Using 3.5 cubic feet of mortar per 100 square feet of wall, determine the amount of mortar needed for the CMU fence in Problem 1.

4. Determine the cost to lay the CMU block in Problem 1. It takes 5.5 mason labor hours and 7.5 laborer labor hours to construct 100 square feet of wall. The average wage rate for the masons is $39.74 per labor hour and the average wage rate for laborers is $22.35 per labor hour.

5. Determine the number of 8-inch-high by 8-inch-wide by 16-inch-long concrete blocks required to complete the wall in Figures 23-2 and 23-3. The overhead doors are 10 feet wide by 12 feet high. If lintel blocks are required wherever the #4 horizontal bars are located and above the doors, how many plain blocks and how many lintel blocks are needed for the wall?

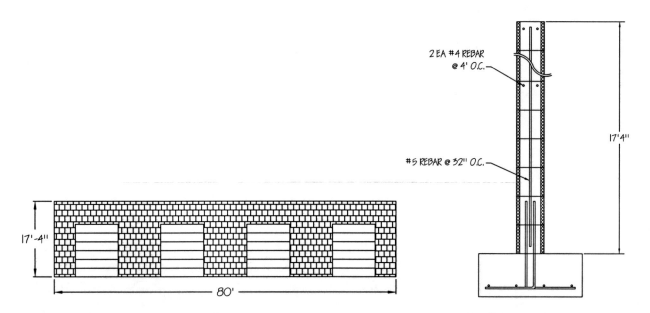

Figure 23-2   Elevation   Figure 23-3   Wall Section

6. Determine the amount of rebar needed for the CMU wall in Problem 5. In addition to the rebar shown in Figure 23-3, there is one #6 rebar located at either side of the overhead doors, which extend from the bottom of the door to 16 inches above the top of the door and two #4 rebar above the doors. Allow for two inches of cover on the vertical bars. Add 8 percent for lap and waste to the horizontal bars and 30 percent for lap and waste to the vertical bars.

7. Using 3.0 cubic feet of mortar per 100 square feet of wall, determine the amount of mortar needed for the CMU wall in Problem 5.

8. Determine the cost to lay the CMU block in Problem 5. It takes 6 mason labor hours and 8.5 laborer labor hours to construct 100 square feet of wall. The average wage rate for the masons is $42.33 per labor hour and the average wage rate for laborers is $21.03 per labor hour.

9. Determine the number of modular bricks needed for the wall in Figure 23-4. Assume that 675 bricks are required for 100 square feet of wall. Add 4 percent for waste.

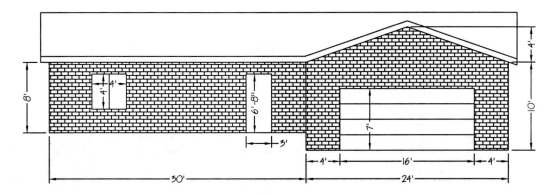

Figure 23-4 Elevation

10. Determine the number of modular bricks needed for the wall in Figure 23-5. The windows are 4 feet high by 4 feet wide and the door is 6 feet wide by 7 feet high. Assume that 675 bricks are required for 100 square feet of wall. Add 5 percent for waste.

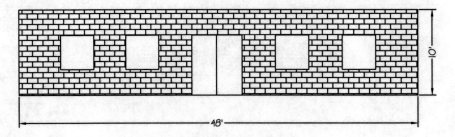

Figure 23-5 Elevation

Exercise 24: Metals

This exercise goes with Chapter 12 of *Estimating in Building Construction*.

1. Prepare a structural steel materials list for the roof-framing plan shown in Figure 24-1. The columns are 16 feet high. How many pounds of steel need to be purchased for the roof?

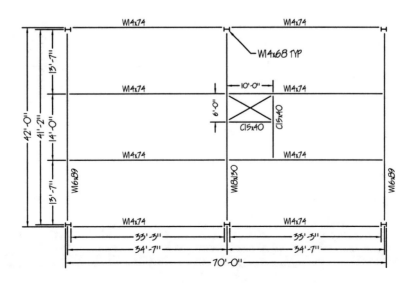

Figure 24-1   Steel Framing Plan

2. Prepare a structural steel materials list for the roof-framing plan shown in Figure 24-2. The columns are 18 feet high and weigh 76 pounds per foot. How many pounds of steel need to be purchased for the roof?

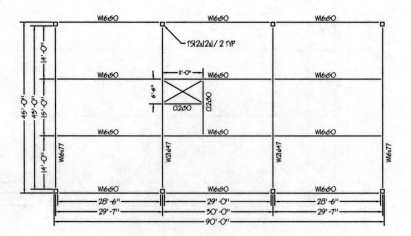

Figure 24-2   Steel Framing Plan

3. Prepare a steel and steel joist materials list for the roof-framing plan shown in Figure 24-3 and determine the number of pounds of joists, beams, and columns that need to be purchased for the roof. The columns are 17 feet high and weigh 76 pounds per foot. The weights for the joist can be found in Table 12-7 of *Estimating in Building Construction.*

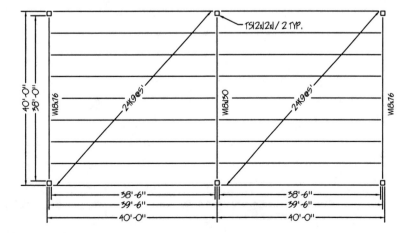

Figure 24-3   Steel Framing Plan

4. Prepare a steel and steel joist materials list for the roof-framing plan shown in Figure 24-4 and determine the number of pounds of joists, beams, and columns that need to be purchased for the roof. The columns are 15 feet high and weigh 58 pounds per foot. The weights for the joist can be found in Table 12-7 of *Estimating in Building Construction*.

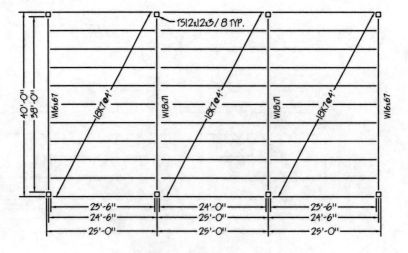

Figure 24-4 Steel Framing Plan

5. Determine the amount of metal deck that needs to be purchased to construct the floor shown in Figure 24-1. Add 12 percent for waste and lapping. Ignore the opening.

6. Determine the cost to install the metal deck for the floor shown in Figure 24-1 using a productivity of 3.1 labor hours per square (100 square feet) and an average labor rate of $37.58 per labor hour.

7. Determine the amount of metal deck that needs to be purchased to construct the floor shown in Figure 24-2. Add 15 percent for waste and lapping. Ignore the opening.

8. Determine the cost to install the metal deck for the floor shown in Figure 24-2 using a productivity of 2.8 labor hours per square (100 square feet) and an average labor rate of $35.11 per labor hour.

Exercise 25: Wood Framed Floors

This exercise goes with Chapter 13 of *Estimating in Building Construction*.

1. Determine the number of board feet in ten 2 × 10s @ 12′.

2. Determine the purchase quantity and number of board feet of 2 × 12 needed to construct the girder in Figure 25-1. The foundation walls are 8 inches thick and the bearing distance for the girder is 6 inches.

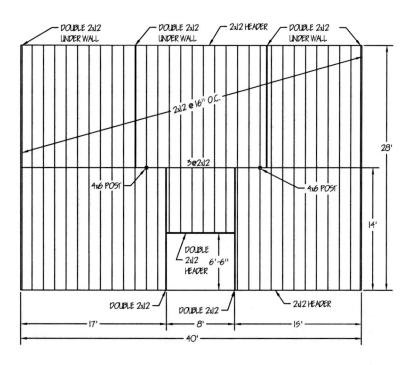

Figure 25-1 Floor Framing Plan

3. A 2 × 4 sill plate runs around the floor in Figure 25-1. Determine the purchase quantity and number of board feet of 2 × 4 needed for the sill.

4. Determine the purchase quantity and number of board feet of 2 × 12 joist needed for the floor in Figure 25-1. The joists are spaced 16 inches on center. Be sure to include the headers.

5. Metal bridging is placed at the center of spans over 10 feet and where the joists pass over the girder. Determine the number of pieces of metal bridging needed for the floor in Figure 25-1.

6. Determine the number of 4-foot by 8-foot sheets of plywood needed for the floor in Figure 25-1. Ignore the opening.

7. Determine the purchase quantity and number of board feet of 2 × 12 needed to construct the girders in Figure 25-2. The foundation walls are 8 inches thick and the bearing distance for the girder is 6 inches.

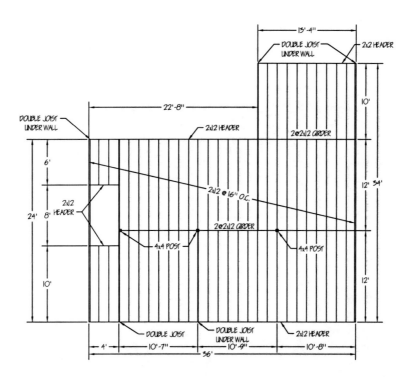

Figure 25-2   Floor Framing Plan

8. A 2 × 4 sill plate runs around the floor in Figure 25-2. Determine the purchase quantity and number of board feet of 2 × 4 needed for the sill.

9. Determine the purchase quantity and number of board feet of 2 × 12 joist needed for the floor in Figure 25-2. The joists are spaced 16 inches on center. Be sure to include the headers.

10. Metal bridging is placed at the center of spans over 10 feet and where the joists pass over the girder. Determine the number of pieces of metal bridging needed for the floor in Figure 25-2.

11. Determine the number of 4-foot by 8-foot sheets of plywood needed for the floor in Figure 25-2. Ignore the opening.

Exercise 26: Wood Framed Walls

This exercise goes with Chapter 13 of *Estimating in Building Construction*.

1. The garage shown in Figure 26-1 is constructed of 2 × 4 lumber and includes a single treated bottom (sole) plate and two untreated top plates. Determine the purchase quantity and number of board feet of treated and untreated lumber needed for the walls.

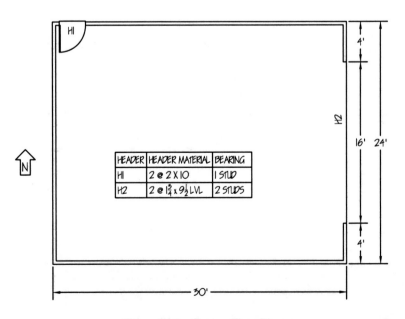

Figure 26-1 Garage Floor Plan

2. Determine the number of 2 × 4 @ 92 5/8" studs needed for the garage in Figures 26-1 and 26-2. The studs are spaced 16 inches on center. Add two studs for each door and corner. Ignore the gable ends and studs over the overhead door.

3. Determine the materials needed for the headers of the garage shown in Figure 26-1. The personnel door is 3 feet wide.

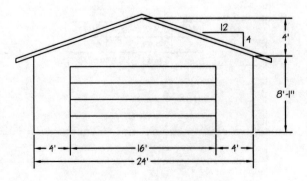

Figure 26-2   East Elevation

4. Determine the number of 8-foot-long studs needed for the gable ends of the garage shown in Figures 26-1 and 26-2. The gable end on the west elevation is the same as the gable end on the east elevation.

5. Determine the number of 4-foot by 8-foot sheets of plywood needed to cover the exterior of the garage in Figures 26-1 and 26-2.

6. The residence shown in Figure 26-3 is constructed of 2 × 4 lumber and includes a single treated bottom (sole) plate and two untreated top plates. Determine the purchase quantity and number of board feet of treated and untreated lumber needed for the walls.

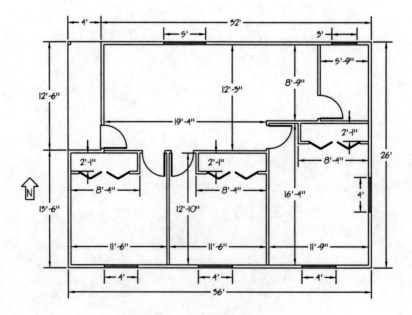

Figure 26-3 Residential Floor Plan

7. Determine the number of 2 × 4 @ 92 5/8" studs needed for the residence in Figures 26-3 and 26-4. The studs are spaced 16 inches on center. Add two studs for each door, window, intersection, and corner. Ignore the gable ends.

8. Determine the materials needed for the headers for the residence shown in Figure 26-3. Two 2 × 10s are used for the headers in the exterior walls and the wall running east-west through the house. The remaining headers are two 2 × 4s. The swinging doors are 3 feet wide and the closet doors are 6 feet wide.

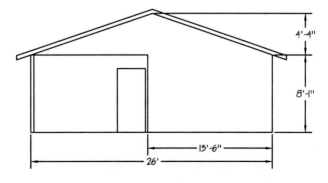

Figure 26-4   West Elevation

9. Determine the number of 8-foot-long studs needed for the gable ends of the residence shown in Figures 26-3 and 26-4. The gable end on the east elevation is the same as the gable end on the west elevation.

10. Determine the number of 4-foot by 8-foot sheets of plywood needed to cover the exterior of the residence in Figures 26-3 and 26-4.

Exercise 27: Wood Framed Roofs

This exercise goes with Chapter 13 of *Estimating in Building Construction*.

1. Determine the purchase quantity of 2 × 6 rafters needed for the roof in Figure 27-1. The slope of the roof is 4:12.

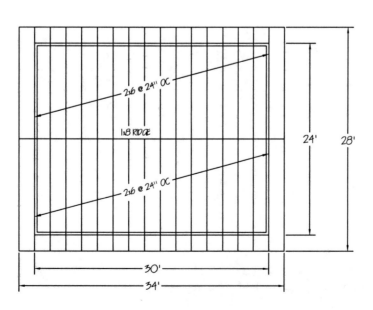

Figure 27-1 Roof Framing Plan

2. Collar ties are used to prevent the rafters of the roof in Figure 27-1 from spreading. The collar ties are 1 × 6s spaced 4 feet on center and are located one-third of the distance from the ridge of the roof to the top of the walls. Determine the purchase quantity of 1 × 6 needed for the collar ties.

3. Determine the purchase quantity of 1 × 8s needed for the ridge of the roof in Figure 27-1.

4. Determine the number of 4-foot by 8-foot sheets of plywood needed to cover the roof in Figure 27-1.

5. How many linear feet of 1 × 8 fascia and 1/2-inch A-C plywood soffit are needed for the roof in figure 27-1?

6. Determine the purchase quantity of 2 × 8 rafters needed for the roof in Figure 27-2. The slope of the roof is 4:12.

7. Collar ties are used to prevent the rafters of the roof in Figure 27-2 from spreading. The collar ties are 1 × 6s spaced 4 feet on center and are located one-third of the distance from the ridge of the roof to the top of the walls. Determine the purchase quantity of 1 × 6 needed for the collar ties.

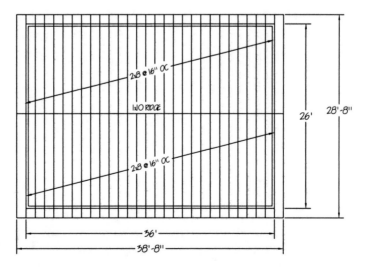

Figure 27-2 Roof Framing Plan

8. Determine the purchase quantity of 1 × 10s needed for the ridge of the roof in Figure 27-2.

9. Determine the number of 4-foot by 8-foot sheets of plywood needed to cover the roof in Figure 27-2.

10. How many linear feet of 1 × 10 fascia and 1/2-inch A-C plywood soffit are needed for the roof in figure 27-2?

Exercise 28: Insulation

This exercise goes with Chapter 14 of *Estimating in Building Construction*.

1. How many rolls of R-19 insulation are required to insulate the floor shown in Figure 28-1? The insulation comes in rolls 15 inches wide by 56 feet long.

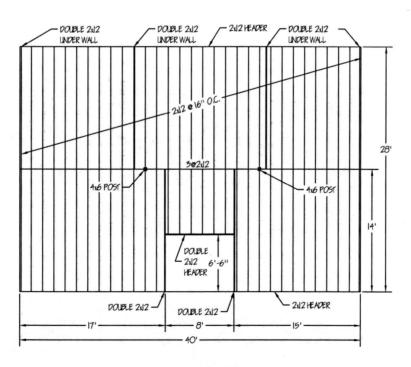

Figure 28-1 Floor Framing Plan

2. How many rolls of R-19 insulation are required to insulate the floor shown in Figure 28-2? The insulation comes in rolls 15 inches wide by 56 feet long.

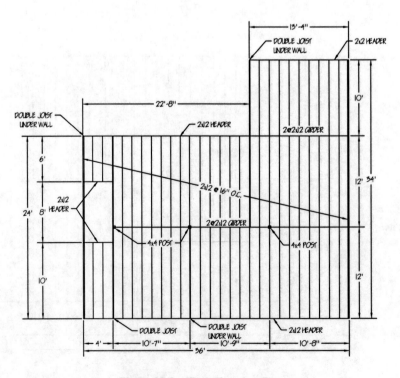

Figure 28-2 Floor Framing Plan

3. How many rolls of R-13 insulation are required to insulate the walls of the building shown in Figures 28-3 and 28-4? The insulation comes in rolls 15 inches wide by 56 feet long. On the gable ends of the building (east and west ends), the insulation is run to the underside of the rafters. Be sure to deduct the area of the garage door, but do not deduct the area of the personnel door. The west side of the building is similar to the east side.

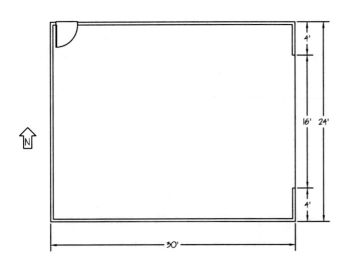

Figure 28-3 Garage Floor Plan

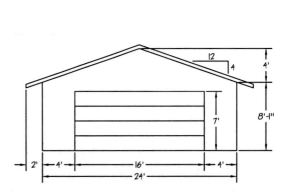

Figure 28-4 East Elevation

4. How many rolls of R-19 insulation are required to insulate the roof of the building shown in Figures 28-3 and 28-4? The insulation comes in rolls 23 inches wide by 56 feet long. The insulation is installed between the rafters, which are spaced 24 inches on center.

5. How many rolls of R-13 insulation are required to insulate the exterior walls of the building shown in Figure 28-5? The wall height is 8 feet 1 inch high. The insulation comes in rolls 15 inches wide by 56 feet long. Do not deduct for the doors or windows.

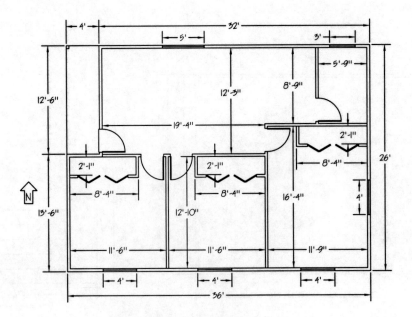

Figure 28-5   Residential Floor Plan

6. How many rolls of R-19 insulation are required to insulate the roof of the building shown in Figure 28-5? The insulation comes in rolls 15 inches wide by 56 feet long. The insulation is installed between the ceiling joists, which are run horizontally and are spaced 16 inches on center.

Exercise 29: Waterproofing and Roofing

This exercise goes with Chapter 14 of *Estimating in Building Construction*.

1. Determine the square feet of waterproofing needed for a rectangular basement 26 feet by 40 feet. The waterproofing is 8 feet high. How many gallons of waterproofing are needed if one gallon of waterproofing will cover 60 square feet?

2. Determine the cost to waterproof the basement in Problem 1 using a productivity of 1.00 labor hour per square (100 square feet) and an average labor rate of $23.17 per labor hour.

3. Determine the square feet of waterproofing needed for the basement shown in Figures 29-1 and 29-2. The waterproofing starts 6 inches from the top of the wall and continues to the top outside corner of the footing. How many gallons of waterproofing are needed if one gallon of waterproofing will cover 65 square feet?

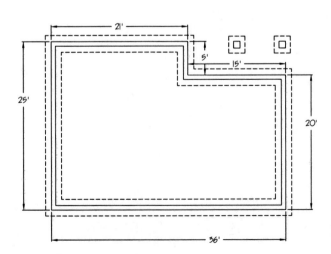

Figure 29-1   Foundation Plan

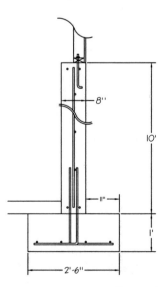

Figure 29-2   Foundation

4. Determine the cost to waterproof the basement shown in Figures 29-1 and 29-2 using a productivity of 0.95 labor hours per square (100 square feet) and an average labor rate of $24.42 per labor hour.

5. How many rolls of felt are needed for the roof shown in Figure 29-3? The slope of the roof is 4:12. The felt comes in rolls 36 inches wide by 67 feet long. The felt is lapped 19 inches.

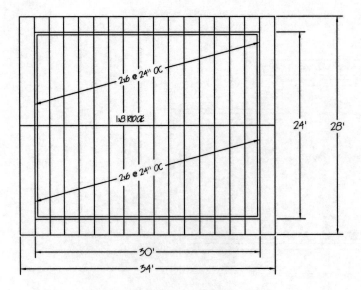

Figure 29-3 Roof Framing Plan

6. How many squares of roofing and how many pieces of ridge shingles are needed for the roof in Figure 29-3? The slope of the roof is 4:12 and the exposure is 5 inches wide.

7. How many rolls of felt are needed for the roof shown in Figure 29-4? The slope of the roof is 5:12. The felt comes in rolls 36 inches wide by 134 feet long. The felt is lapped 6 inches.

8. How many squares of roofing and how many pieces of ridge shingles are needed for the roof in Figure 29-4? The slope of the roof is 5:12 and the exposure is 5 inches wide.

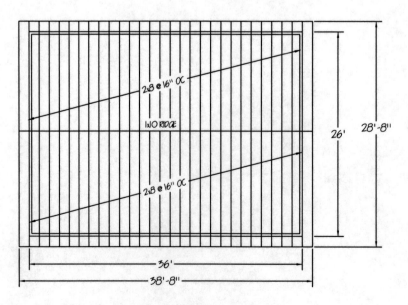

Figure 29-4   Roof Framing Plan

Exercise 30: Doors and Windows

This exercise goes with Chapter 15 of *Estimating in Building Construction*.

1. Prepare a materials list for the residential windows in Figure 30-1.

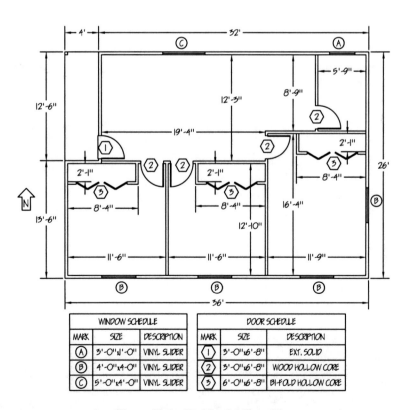

Figure 30-1   Residential Floor Plan

2. Determine the cost to install the windows in Figure 30-1 using a productivity of 3.0 labor hours per window and an average labor rate of $29.50 per labor hour.

3. Prepare a materials list for the residential doors in Figure 30-1. Be sure to specify swing.

4. Determine the cost to install the doors in Figure 30-1 using the productivity rates in Table 30-1 and an average labor rate of $29.50 per labor hour.

Table 30-1 Productivity	
Type	Labor-hours per door
Solid doors, swinging	2.5
Hollow core, swinging	2.0
Hollow core, bifold	3.0

5. Determine the lineal feet of aluminum tubing needed for the storefront in Figure 30-2.

6. Prepare a materials list for the glass required for the storefront in Figure 30-2. How many square feet of glass is required for the storefront?

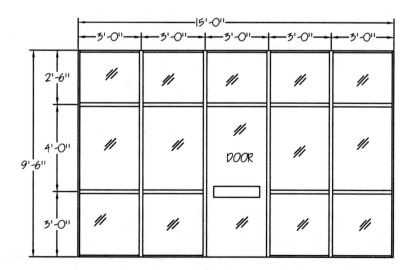

Figure 30-2 Storefront Elevation

7. Determine the cost to install the storefront in Figure 30-2 using a productivity of 0.08 labor hours per square foot and an average labor rate of $32.96 per labor hour.

Exercise 31: Drywall and Plaster

This exercise goes with Chapter 16 of *Estimating in Building Construction*.

1. Determine the number of steel studs needed for the tenant finish shown in Figure 31-1. The exterior walls are existing. The studs are spaced 16 inches on center. Add two studs for each door and each intersection.

2. How many 20-foot-long pieces of runner track are needed for the tenant finish in Figure 31-1?

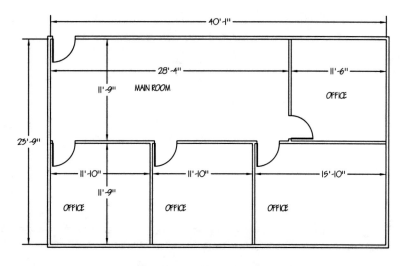

Figure 31-1   Office Floor Plan

131

3. Determine the number of 4-foot-wide by 8-foot-high sheets of drywall needed for the tenant finish in Figure 31-1. The walls are 8 feet high. The exterior walls need drywall on the interior side only. The interior walls need drywall on both sides. Add 5 percent waste.

4. Determine the number of steel studs needed for the tenant finish shown in Figure 31-2. The exterior walls are existing. The studs are spaced 16 inches on center. Add two studs for each door and each intersection.

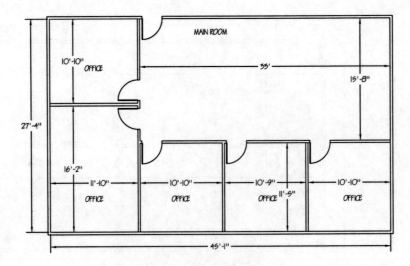

Figure 31-2 Office Floor Plan

5. How many 20-foot-long pieces of runner track are needed for the tenant finish in Figure 31-2?

6. Determine the number of 4-foot-wide by 8-foot-high sheets of drywall needed for the tenant finish in Figure 31-2. The walls are 8 feet high. The exterior walls need drywall on the interior side only. The interior walls need drywall on both sides. Add 5 percent waste.

7. Determine the number of 4-foot-wide by 8-foot-high sheets of drywall needed for the walls of the garage in Figure 31-3. The walls are 8 feet high and the overhead door is 7 feet high by 16 feet wide. Only the inside walls of the garage need drywall.

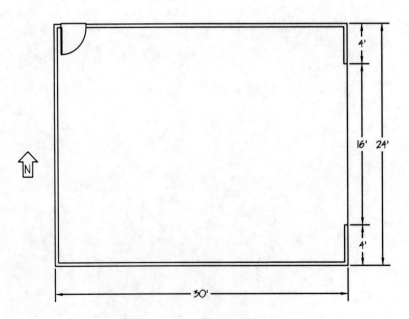

Figure 31-3 Garage Floor Plan

8. Determine the number of 4-foot-wide by 8-foot-high sheets of drywall needed for the ceiling of the garage in Figure 31-3. The ceiling is flat. Add 5 percent waste.

9. Determine the number of 4-foot-wide by 8-foot-high sheets of drywall needed for the walls of the residence in Figure 31-4. The walls are 8 feet high. The exterior walls get drywall only on the inside, and the inside walls get drywall on both sides. Add 5 percent waste.

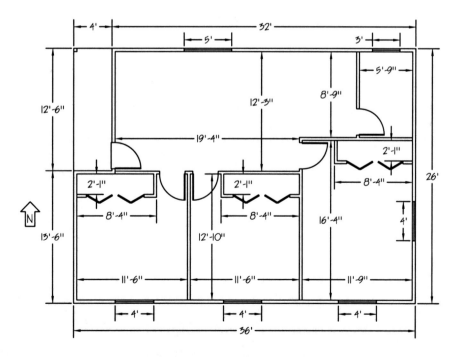

Figure 31-4 Residential Floor Plan

10. Determine the number of 4-foot-wide by 8-foot-high sheets of drywall needed for the ceiling of the residence in Figure 31-4. The ceiling is flat. Add 5 percent waste.

11. The walls of the residence shown in Figure 31-4 are plastered with a 1/4 inch of plaster over metal lath. The exterior walls get plaster only on the inside, and the inside walls get plaster on both sides. The walls are 8 feet high. Determine the number of cubic feet of plaster needed for the walls.

12. The ceiling of the residence shown in Figure 31-4 is plastered with 3/8 of an inch of plaster over metal lath. Determine the number of cubic feet of plaster needed for the ceiling.

Exercise 32: Flooring, Tile, and Paint

This exercise goes with Chapter 16 of *Estimating in Building Construction*.

1. Determine the number of square feet of resilient flooring needed for the residence in Figure 32-1. Add 5 percent for waste. How many gallons of adhesive are needed using a coverage of one gallon per 120 square feet of floor?

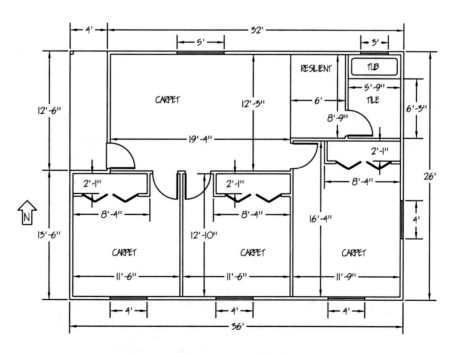

Figure 32-1 Residential Floor Plan

2. Determine the number of square yards of carpet needed for the residence in Figure 32-1. The carpet is 12 feet wide.

3. How many gallons of primer and paint are needed to paint the walls of the residence shown in Figure 32-1? The walls are to receive one coat of primer and two coats of paint. The exterior walls only need paint on the interior sides. The walls are 8 feet high. One gallon of primer will cover 275 square feet and one gallon of paint will cover 350 square feet with a single coat. The swinging doors are 3 feet wide by 6 feet 8 inches high, the closet doors are 6 feet wide by 6 feet 8 inches high, and the windows are 4 feet high.

4. How many gallons of primer and paint are needed to paint the ceiling of the residence shown in Figure 32-1? The ceiling is to receive one coat of primer and two coats of paint. One gallon of primer will cover 275 square feet and one gallon of paint will cover 350 square feet with a single coat.

5. How many square feet of tile are needed for floors of the residence in Figure 32-1?

6. Determine the number of square feet of resilient flooring needed for the office building in Figure 32-2. Add 5 percent for waste. How many gallons of adhesive are needed using a coverage of one gallon per 150 square feet of floor?

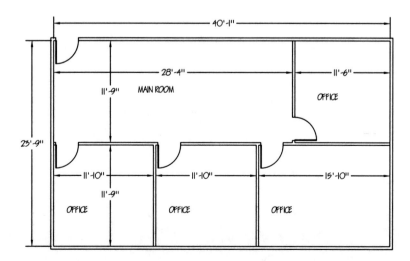

Figure 32-2 Office Floor Plan

7. How many lineal feet of 4-inch-wide rubber base is needed for the office building in Figure 32-2? The doors are 3 feet wide and base is not needed on the outside of the exterior walls. Add 5 percent for waste. How many gallons of adhesive are needed using a coverage of one gallon per 150 square feet of base?

8. Determine the number of square yards of carpet needed for the office building in Figure 32-2. The carpet is 12 feet wide.

9. How many gallons of primer and paint are needed to paint the walls of the office shown in Figure 32-2? The walls are to receive one coat of primer and two coats of paint. The exterior walls only need paint on the interior sides. The walls are 8 feet high. One gallon of primer will cover 300 square feet and one gallon of paint will cover 400 square feet with a single coat. The doors are 3 feet wide by 6 feet 8 inches high.

10. How many square feet of tile are needed for walls and floor of the bathroom in Figure 32-3? The wall tile is 4 feet high and the doors are 3 feet wide.

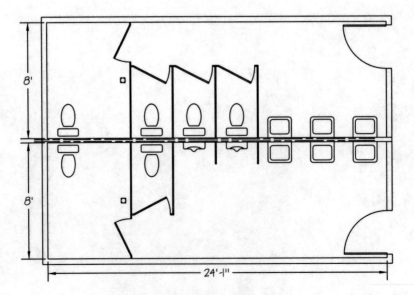

Figure 32-3   Bathroom Floor Plan

Exercise 33: Electrical

This exercise goes with Chapter 17 of *Estimating in Building Construction*.

1. Determine the estimated cost of the electrical for the residence shown in Figure 33-1. A subcontractor has given you the following unit prices for wiring a residence. The associated wiring is included in the prices of each of the items. What is the fixture allowance for the residence?

    Switches = $50
    Outlets = $50
    GFI outlets = $60
    Light fixtures (includes $30 fixture allowance) = $80
    Main service = $1,000

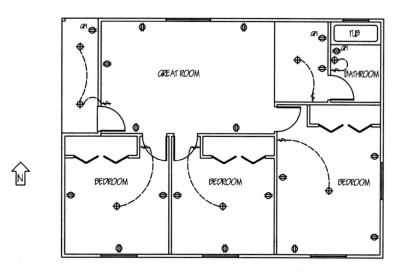

Figure 33-1 Electrical Plan

2. Write a scope of work for the electrical of the residence shown in Figure 33-1.

Exercise 34: Plumbing

This exercise goes with Chapter 18 of *Estimating in Building Construction*.

1. Determine the estimated cost to replace the plumbing fixtures in the bathroom shown in Figure 34-1. A subcontractor has given you the following unit prices to replace the fixtures.

    Lavatory (sink) = $695
    Water closet (toilet) = $610
    Urinal = $820

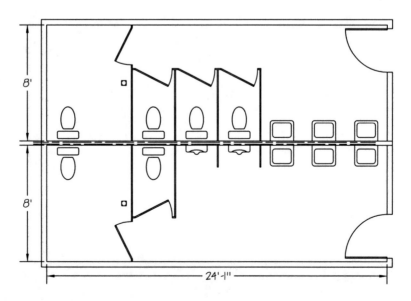

Figure 34-1   Bathroom Floor Plan

2. Write a scope of work for the plumbing of the building shown in Figure 34-1.

Exercise 35: Heating, Ventilating, and Air-Conditioning

This exercise goes with Chapter 19 of *Estimating in Building Construction*.

1. Determine the estimated cost for the heating system for the residence shown in Figure 35-1. A subcontractor has given you a cost of $1,900 for the heating unit and a cost of $195 per register, which includes the ductwork from the heating unit to the register. A register is required in each of the rooms. Rooms over 150 square feet require one register per 150 square feet or fraction thereof.

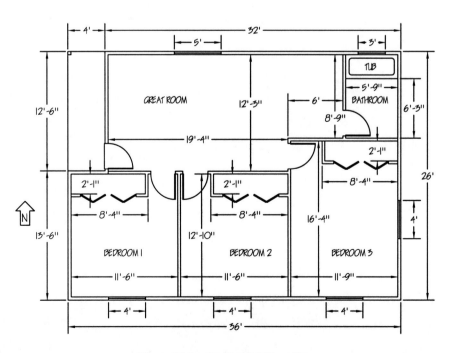

Figure 35-1 Residential Floor Plan

147

2. Write a scope of work for the HVAC of the residence shown in Figure 35-1.

Exercise 36: Profit

This exercise goes with Chapter 20 of *Estimating in Building Construction*.

1. What should an estimator take into account when setting the profit margin for a job?

2. Your company is bidding on a religious building. You estimated that its construction costs are $1,865,400, which includes project overhead. Your company marks up all construction costs 5 percent to cover main-office overhead. Your boss wants to make a 3 percent profit on the construction costs (excluding main-office overhead). What is your company's bid for the project?

3. Your company is bidding on an office building. The estimated construction costs are as follows:

   Labor = $985,466
   Materials = $1,091,256
   Equipment = $100,754
   Subcontractor's bids = $6,254,800

   Your company marks up labor and materials 15 percent, equipment 20 percent, and subcontractor's bids 4 percent to cover main-office overhead and profit. What is your company's bid for the project?

Exercise 37: Other Estimating Methods

This exercise goes with Chapter 21 of *Estimating in Building Construction*.

1. Using the project comparison method, prepare an estimate for a two-story home in your local area. The home is to have three bedrooms, two bathrooms, a living room, kitchen and dining facilities, and a two-car garage. The square footage should be between 1,800 and 2,200 square feet. Exclude the lot, all site work, and all permits from your estimate. To prepare this estimate, you will need to find two or three projects to use as a comparison. These may be obtained via the Internet or talking to local builders. Be sure to adjust the prices as necessary to make the projects truly comparable, including adjustment for age and quality of materials. Be prepared to share the projects you used as a comparison and your calculations with the class, if your instructor chooses to do so.

2. Last year your construction company built a 2,400 square foot home with a two-car garage for $255,658. Another client wants a similar home built, except they want a three-car garage. It is estimated that the garage will cost an additional $9,500 and that costs have risen 5 percent during the last year. Using this information, prepare a preliminary estimate for the house.

3. Your company has been asked to prepare a preliminary cost estimate for a 10,000 square foot automotive repair garage, with four hydraulic lifts. From past projects, you have determined that the average cost per square foot to construct an automotive repair garage is $107.60 and excludes the cost of the hydraulic lifts. The lifts cost $16,000 each. Ignoring inflation, prepare an estimate for the repair garage.

4. Your company has been asked to prepare a preliminary cost estimate for a three-story, 22,000 square foot office building. From projects you have constructed during the last year, you have determined that the average cost per square foot to construct an office building is $158.59. Using an inflation rate of 5 percent, prepare an estimate for the office building.

5. List the items that you would need to include in a concrete footing and foundation wall assembly.

6. Using Figures 21-2 and 21-3 from *Estimating in Building Construction*, determine the cost for a 250-foot by 200-foot warehouse. The exterior walls are to be concrete block, which are used as bearing walls. The warehouse is 20 feet high. Include ten 7-foot by 8-foot dock levelers in the costs.

7. Using Figures 21-2 and 21-3 from *Estimating in Building Construction*, determine the cost for a 125-foot by 200-foot warehouse. The exterior walls are to be galvanized steel siding over a steel frame. The warehouse is 26 feet high. Include 50 feet of asphalt around the building in your estimate.